There are many v
with our steps . .

1

or hand to hand.

Measuring tapes and number charts tell us how tall we stand.

How heavy is the food?
A scale will read the weight.

Watches and clocks can measure time so we know we won't be late.

A doctor uses special tools
to measure how we grow.

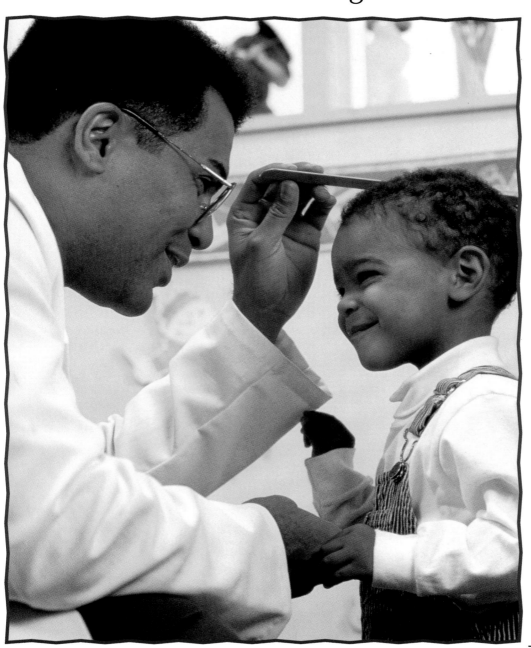

Thermometers **tell if your temperature's high,**

and measure
the cold when
it snows.

How will they measure
the sugar they need?

With a measuring cup?

Or a scale?

Which tool would you use to measure the length of a dog, from nose to tail?

15 16 17 18 19 20 21 22 23 24 25 26 27 28

13

Measuring tools will let us know how hot, how heavy, how wide.

How long is it? How tall is it? How much water is inside?

There are lots of things to measure,
so come along with me!
What things will you measure
with all the tools you see?